W9-DDP-344

MOUNTAINS

Alan Collinson

DILLON PRESS
NEW YORK

First American publication 1992 by Dillon Press, Macmillan Publishing Company, 866 Third Avenue, New York, NY 10022

Macmillan Publishing Company is part of the Maxwell Communication Group of Companies

First published by Evans Brothers Limited, 2A Portman Mansions, Chiltern Street, London W1M 1LE

Typeset in England by Fleetlines Typesetters, Southend-on-Sea
Printed in Spain by GRAFO, S.A.—Bilbao

10 9 8 7 6 5 4 3 2 1

Collinson, Alan.
Mountains / Alan Collinson.
p. cm. — (Ecology watch)
Includes index.
Summary: Discusses how mountains are formed, how plants and animals adapt to life in them, and how they influence weather and geographical changes.
ISBN 0-87518-493-6
1. Mountain ecology—Juvenile literature. 2. Mountains—Juvenile literature. [1. Mountains. 2. Mountain ecology. 3. Ecology.]
I. Title. II. Series.
QH541.5.M65C66 1992
574.5'264—dc20 91-34171

Acknowledgments

Editor: Su Swallow
Design: Neil Sayer
Production: Jenny Mulvanny

Illustrations: David Gardner, Graeme Chambers
Maps: Hardlines, Charlbury

For permission to reproduce copyright material the author and publishers gratefully acknowledge the following:

Cover (Llamas, Peru) W.E. Townsend, Jr., Bruce Coleman Limited
Title page (Children collecting bedding, Bhutan) Mark Edwards, Still Pictures

p4 (top) Chelmick/ECOSCENE, (bottom) M. Timothy O'Keefe, Bruce Coleman Limited **p6** Dennis Green, Bruce Coleman Limited **p7** John Waters, Bruce Coleman Limited **p8** Keith Gunnar, Bruce Coleman Limited **p9** (top) Keith Gunnar, Bruce Coleman Limited, (bottom) Gunter Ziesler, Bruce Coleman Limited **p11** Stephen J. Krasemann, Bruce Coleman Limited **p12** (top left) Zig Leszczynski, Oxford Scientific Films, (top right) Worldwide Fund for Nature, Bruce Coleman Limited, (bottom) Hans Reinhard, Bruce Coleman Limited **p13** Hans Reinhard, Bruce Coleman Limited **p14** Martin Dohrn, Bruce Coleman Limited **p15** (top) Tom Leach, Oxford Scientific Films, (middle) Roland Mayr, Oxford Scientific Films, (bottom) Hans Reinhard, Bruce Coleman Limited **p17** Jane Burton, Bruce Coleman Limited, (inset top) David Purslow/ECOSCENE, (inset bottom) Laurie Campbell, NHPA **p18** Hans Reinhard, Bruce Coleman Limited **p19** Dr Eckart Pott, Bruce Coleman Limited **p20** Ben Osborne, Oxford Scientific Films **p21** Worldwide Fund for Nature, Bruce Coleman Limited **p22** John Shaw, NHPA **p23** Sally Morgan/ECOSCENE **p24** (left) Brown/ECOSCENE, (right) Dr R. Parks, Oxford Scientific Films **p25** Jutta Hösel, NHPA **p26** Keith Gunnar, Bruce Coleman Limited **p27** (top) Jeff Foott Productions, Bruce Coleman Limited, (middle) Sally Morgan/ECOSCENE, (bottom) Erwin and Peggy Bauer, Bruce Coleman Limited **p28** (top) Erwin and Peggy Bauer, Bruce Coleman Limited, (bottom) Jeff Foott, Bruce Coleman Limited **p29** John Shaw, Bruce Coleman Limited **p30** Gunter Ziesler, Bruce Coleman Limited **p31** G.I. Bernard, NHPA, (inset) W.E. Townsend, Jr., Bruce Coleman Limited **p32** (left) Gunter Ziesler, Bruce Coleman Limited, (right) Michael Fogden, Oxford Scientific Films **p33** N.A. Callow, NHPA **p34** Dieter and Mary Plage, Bruce Coleman Limited **p35** Michael Dick/Animals Animals, Oxford Scientific Films **p37** (top and bottom) Richard Packwood, Oxford Scientific Films, (middle) Andrew Plumptre, Oxford Scientific Films **p38** (top) David Simonson, Oxford Scientific Films, (bottom) Thomas Buchholz, Bruce Coleman Limited **p39** Hawkes/ECOSCENE **p41** Corbett/ECOSCENE **p42** (top) Mark Carwadine, Biotica, (bottom) E. Hanumantha Rao, NHPA **p43** E. Lauber, Oxford Scientific Films **p44** L.C. Marigo, Bruce Coleman Limited.

Contents

Introduction

What is a mountain?

Some mountains, such as the Rockies, the Alps, or the Himalayas, are very high, cold, and often have sharp peaks. They may be covered in forests or even be too high for trees or other plants to grow. The highest areas in places such as Scotland or New Jersey, may also be called mountains, but these mountains are very different from the Rockies, Alps, and Himalayas. They are not very high, have no sharp peaks and are not especially cold. The Watchung Mountains in New Jersey, for example, are less than 500 feet high. The highest parts of the Scottish mountains are about 4,000 feet high, yet a 13,000-foot peak in Nepal in the Himalayas is regarded by local people as only "foothills."

In the past, mountains in cooler parts of the world attracted few people to live on them. The steep slopes, thin soils, and harsh climate make farming more difficult than on the surrounding lowlands. This is true of mountains in Europe, China, central Asia, and North America. In parts of the world with hot climates, the cooler high land of the mountains offered a pleasant alternative to the hot lowlands and often supported large numbers of people.

△ Spring gentian in the Alps

▽ A family on the move in the Andes.

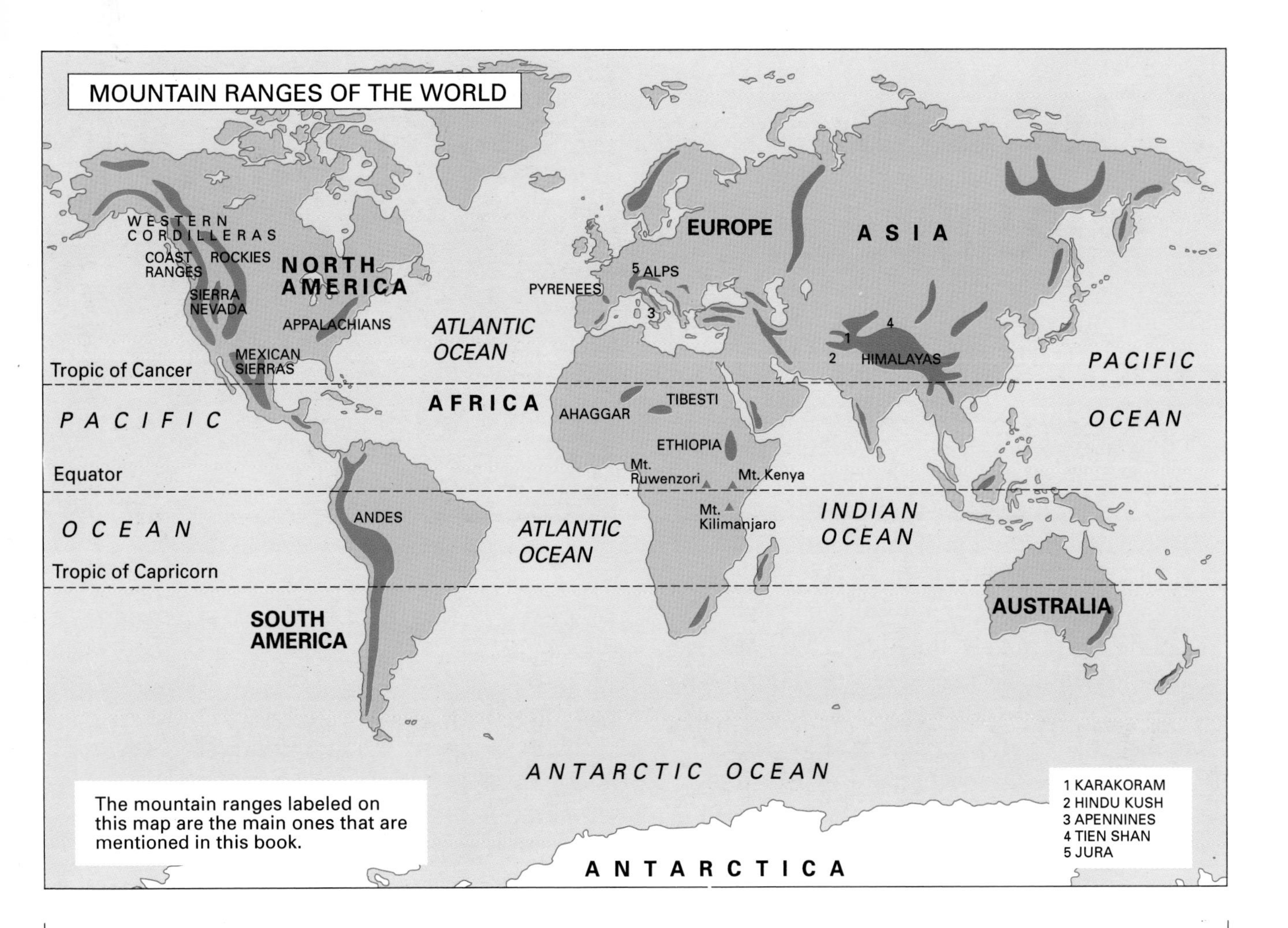

The delicate heights

Over the last 150 years, new railroads and roads have made mountains in both cool and hot climates much more accessible. More and more people now visit them. Some come for vacations, skiing, or mountain climbing. Some come to dig mines and quarries. Some build dams for water storage and power production, and yet others cut timber. There are now more people living and working in mountains than ever before.

The new activities and the extra people are affecting the plants and animals that live in the mountains. Of all the world's environments, mountains are possibly the most easily damaged by these kinds of activities. Mountain wildlife is already coping with extremes of climate. Any pressures from people are therefore liable to threaten its fragile existence.

People can also suffer when mountains are mistreated, even if they do not live there permanently. For example, many mountains have high rainfall. The strong rush of water through the thin soil does not usually wash the soil away, because trees and other plants keep it in place. The plants also soak up some of the water. If the trees are cut down in large numbers (for lumber, or to clear land for mining or skiing, for example), the soil is carried away. Rivers become choked with sand and gravel, and flooding may be widespread. This has happened on a large scale in the United States, India, China, and Southeast Asia.

Fortunately, there are now very few countries with mountains that are not conducting an "ecology watch" on the resources of their mountain regions. New forests are being planted, so that soil erosion and flooding can be controlled, and plant and animal populations protected. Provided that long-term conservation is seen as more important than short-term gain, mountain environments and their wildlife should survive.

Words printed in **bold** are explained at the end of each section.

Mountain climate

△ Heather can survive on cold mountain slopes.

▷ Rice terraces face the sun on a mountainside in the Philippines.

Height and heat

All living things are affected by heat conditions. Different species of plants like different temperatures, but there are upper and lower limits when they cease to be active in growth. If the temperature falls below 40°F for a period of more than a few days, plants cannot grow, although they do not die. Fortunately, the upper safe limit for heat (around 104°F) is very rare, although it can occur near volcanoes or on desert rocks.

Animals can remain active at lower and higher temperatures than these, provided they can control their internal heat. Warm-blooded animals use insulation such as fur or feathers, and may hibernate to avoid extremes of temperature. Cold-blooded animals such as fish, insects, **reptiles**, and **amphibians** tend to avoid the very cold heights altogether. In mountains, temperature is the most important influence on the ecology of plants and animals.

As height increases, the air gets thinner, drier, and cleaner. Heat given off by the earth in high mountains passes straight through the thin air to outer space, without warming the air at all. Mountains are like open windows letting the heat out. The temperature of the air falls steadily as the land gets higher. A general figure for the fall is about 1.2°F for every 330 feet of height.

Plant patterns

As height increases, plants are sorted out by the increasing cold. In most of the world's higher mountains, different zones of plants follow one after another. The plants in each zone are better adapted to the cold than the ones below them. The plant zones in most mountain areas are also zones of animal distribution. The animals are sorted out not so much by the cold but by the possibilities of getting the right kinds of food plants.

Plants living in high mountains often have to cope with very sudden changes of temperature, from day to night, for example, and from summer to winter. Experiments on heather and pine in the Alps show that even though temperatures are quite high in summer, the plants retain a resistance to cold right through the season. At the time of the first frost in September, they can increase their cold resistance so efficiently that it would take temperatures as low as −44°F to kill their leaves. This is a much greater safety margin than lowland plants have.

On the right side

The shape of the land, as well as the height, affects temperature conditions for plants. In mountains or even low hills, cold air can slide into the bottoms of valleys from the sides. That is because cold air is heavier than warm air. The cold air accumulates in the valley bottoms and can give frosts or cool, foggy conditions. Plant growth can be reduced by the low temperatures, and some plants may be kept out altogether. Farmers living in mountains may not plant certain crops in the valley bottom, for fear of frost damage. People may even build their villages on the mountainside rather than on the valley floor, and build terraces to make flat field surfaces.

Temperature conditions can also vary widely from one side of a valley to another. Sunny sides often have much better conditions for plant growth than shady sides. Slopes facing west and east can be very different in temperature. Usually, west-facing slopes are warmer in winter than east-facing slopes. In the summer it is the other way around. Plants that grow on these slopes may reflect these differences in temperature.

Height and light

Plants use the energy in the white light from the sun to make their food. This process is called photosynthesis. Plants take in carbon dioxide from the air and hydrogen from water to make sugars and starch (carbohydrates). These in turn can be used to make proteins, fats, and new living material. However, these chemicals are destroyed if they are exposed to another kind of light, ultraviolet, which has much more energy than white light.

Fortunately, most ultraviolet (UV) light is absorbed by the atmosphere; at low levels the danger from it is small. However, in mountains the air is so thin there is lots of UV light. The amount also increases with height the nearer the mountains are to the equator. Above the tree zone in the high Himalayas, the Andes, or the high African mountains there is seven times more UV radiation than at sea level in the same **latitude**.

The UV light reduces the growth of leaves and slows down photosynthesis. Mountain plants have to have special chemistry to survive at high levels. This can be seen by comparing plants that grow in the Arctic with the same species from high mountains. The plants look exactly the same, but tests have shown they are different. Plants from mountains in Venezuela, Peru, and Hawaii had leaves that stopped 98 percent of the UV light from getting into the leaf interior. The same plant species from the Arctic were badly damaged by the light. The mountain plants had built-in "sun tan" screens. (See also page 44.)

Volcanic mountains

Many mountain ranges have been created by movement of the earth's crust. These mountains often include active volcanoes, which lie above breaks in the earth's crust. When the volcanoes erupt they present a hazard to people, animals, and plants living in the mountain areas.
Mount St. Helens, in the state of Washington, erupted violently in 1987. Much of the forest around it was burned or buried by the ash and many animals were killed.

△ The Valley of Fire in North America, part of the desert landscape on the leeward side of the Sierra Nevada.
◁ Desert in Patagonia, on the leeward side of the Andes.

Rain and snow

When winds with rain clouds reach mountains and begin to climb, the cold air causes the water vapor to condense and fall as rain. In the highest mountains, the winds eventually dry out as they cross the highest parts. As the wind descends again it warms up, so the cloud droplets begin to evaporate and the rainfall is less. Where the winds come more frequently from one direction than another, the two sides of a mountain range may have quite different kinds of weather: a wet windward side and a drier leeward side, called a "rain shadow." Sometimes, the rain shadow is so strong that it creates a desert on the leeward side of mountains. This happens in South America in Patagonia, and in North America in the western Sierras. The vegetation growing on the wet and dry sides will be very different in such places.

In dry lands, local mountains may be the only sites where rain is likely to fall regularly, and this is where plants can grow. For example, in the Sahara Desert the Ahaggar and Tibesti Highlands are oases of watered land in the middle of desolation. In China, the interior desert basins have well-watered mountains surrounding them. In Peru, the fertile valleys of the coastal deserts are watered from mountain rainfall.

Because rainfall can increase dramatically in mountains, and because slopes are steep and soils thin, the run-off from rain can cause great damage to plants. Landslides of soaked mud and rock can sweep away whole areas of forest. This happens especially where people have cut down some forest, leaving the soil unprotected. Landslides are also frequent where fires have occurred.

In the steeper, higher parts of high mountains, snow causes damage when avalanches occur. These can be triggered off by local warm winds, which are often a part of the local mountain climate. Some of these winds have special names, for example the Chinook in Canada, the Föhn in the Alps, and the Zonda in Argentina.

Winds, in any case, are much stronger in mountain areas than in lowlands. Extra speed may also occur where winds are forced through narrow gaps between mountain ranges. The mistral wind in southern France, for example, blows down the Rhône valley between the Central Massif and the Alps. In Britain, the highest wind speeds recorded are on the tops of the Scottish mountains. In the United States, one of the highest wind speeds ever recorded on land was at Mount Washington in New Hampshire. In one 24-hour period the average wind speed was 130 mph and one gust reached 231 mph. Mount Washington is only just over 6,200 feet. Clearly, any plant that did not hug the ground would be broken very quickly with winds like these. In any case, it is not simply the speed that causes the damage. It is the increased force (push) the wind has with increased speed. (When the wind speed doubles, the force it has increases by four times.) The strong winds dry the soil more, and this may also prevent the growth of trees.

In some cases, the frequent snow on mountains in middle latitudes can be beneficial to plants and animals. If snow falls regularly and does not melt before a good accumulation takes place, the soil is often kept warmer than it would be without the snow cover. Small animals and plants can survive under its protection. Snow acts like a thick insulating blanket because it contains so much air. Where snow comes infrequently, or melts off before the next cover, the soil temperature falls, and plant roots—especially tree roots—can be damaged by frost heaving up the ground. In some places, such as Britain and Norway, frost damage and wind, rather than the cold air alone, tend to control the upper limit of tree growth.

Icy landscapes

About two million years ago the earth's climate suddenly became much colder, and ice began to spread in the mountains. Soon it flowed onto the plains in northern Eurasia, North America, southern Argentina, southeast Australia, New Zealand, and Iceland as great ice sheets. Ice covered much of the land surfaces in these areas. Eventually the ice melted away, only to return again and again until between 12,000 and 10,000 years ago it began to melt away once more. This whole period is called the Ice Age. The glaciers that formed during that time helped to shape most of the world's high mountains.

The glaciers of today (except those on Antarctica and Greenland) are the remnants of these great ice masses. They advance and retreat as the climate gets slightly colder or warmer. A hundred years ago they were advancing; now they are retreating. Near the ice edge, in high lands with severe winters and short summers, the ecology of the plants and animals that live there is like that of the Ice Age itself.

Some mountains also preserve kinds of plants that have disappeared from elsewhere. The mountains become refuges where plants can find exactly the right conditions in the wide variety of climate and soils contained in the mountain mass. In the Himalayas, for example, some warm temperate forests are very like the forests that covered much of western Europe and eastern Asia before the Ice Age. In some of these forests, animals that lost their homes as the forests were destroyed are also preserved. The giant panda is one example (see pages 12 and 36).

▷ Glacier Bay, Alaska. Glacier ice wears the land into U-shaped valleys as it moves slowly downhill.

The giant panda (left) feeds on bamboo in the dense forests high in the mountains of southwest China (above). The bamboo grows abundantly here because of the high rainfall.

This beautiful animal is now threatened by the clearing of bamboo forests for farming as the population of China increases. The threatened panda has been adopted as the symbol for the World Wide Fund for Nature, an organization dedicated to saving wildlife.

Mountain soils

Most soils in the mountains are thin and stony, and there may be large areas of bare rock without soils. Where steep slopes and high rainfall occur, soil is easily washed away and carried off by the rivers (see diagram on page 40). Pieces of broken rock fall down the slopes in heaps called screes. It is very difficult for plants to take root on screes, since they are constantly moving. Only some specialized plants can grow on the loose rock. Rocks are broken up mostly by the action of frost. Water in cracks in the rocks expands when it freezes, which causes rock fragments to snap off. This can happen in even quite dry mountains where dew settles in the cracks.

Because the climate in mountains is colder, plant remains may not rot away easily. In parts where the drainage is poor, they can build up on the surface and form layers of peat. In Britain and Ireland, for example, the tops of the highest mountains may have layers of peat, called blanket peat,

▷ Alpine meadows provide good pasture.

▽ Bilberry grows mainly on moors and peat bogs on high ground.

almost 10 feet thick. At the bottom of this peat layer it is possible to find the remains of oak trees that grew there before the peat formed. The trees were killed as the climate became wetter about 6,000 years ago. Most of the peat is made up of the remains of plants that grow there, such as heather, bilberry, and especially a moss called sphagnum. This plant has many spaces in its cells that hold water, so it acts like a sponge and soaks up the rain.

Acid attack

In most mountains where the climate is wet and cold, the water in the soil becomes very acid. Acid is released from plant remains because there are no insects, worms, bacteria, or fungi to feed on the plants and break them down properly. Minerals that plants need as food such as nitrogen, phosphorus, calcium, iron, and potassium are leached (washed) out of the acid soil easily. Leaching is one of the main reasons that farming on high land is so difficult.

Recently, many of the mountain soils in the eastern United States and northern and eastern Europe have become even more acid. Rivers and lakes have also been affected. The extra acid is being added by rain, which is washing industrial gases such as sulphur dioxide out of the air. The gases are put into the air by the burning of fossil fuels. The acid rain kills leaves and buds, and poisons roots. It also kills fish.

Good grazing

In spite of the poor soil on mountains many plants have become adapted to living on them. On the poorest soils such as the peats, there are only a few species, but where the drainage is good there can be many species. For instance, above the tree line in the Alps, Himalayas, and Rockies there are rich grassy pastures with good drainage. Many kinds of flowers grow in these pastures in spring and summer. The soil is dug and turned over by small burrowing animals such as voles. Grazing animals such as the ibex, yak, mountain goat, deer, buffalo, cattle, and sheep produce good manure, which fertilizes the soil. In all mountains, these pastures above the tree line are named the alpine zone, after the most famous of them in the Alps. The plants that grow there are called alpine-arctic species. The Alps themselves are named after these high pastures, which the local people call *alpen*.

reptiles—cold-blooded animals with backbones and scaly skins, such as snakes and lizards.
amphibians—animals that live mostly on land but breed in water.
latitude—the distance north or south of the equator, measured in degrees at the center of the earth. (One degree latitude = 69.6 miles.)

The mountains of Europe

Forest cover

All of Europe's mountains are well-watered, so forests can grow to quite high levels. In areas like those around the Mediterranean Sea, many of the forests of pine, laurel, oak, cedar, and cypress have been cut or burned down long ago. The soil has often been washed away from the steeper slopes, so there is lots of bare rock with scrubby plants, known as maquis, the French word for this kind of vegetation. In the summer these areas look almost desertlike. In the winter and spring they are much greener.

In most of the other mountain areas of Europe, the lower slopes are covered with **deciduous hardwood** trees such as oak, beech, sweet chestnut, lime, hornbeam, and ash. These trees are mostly below 2,600 feet. As the land gets higher there are forests of pine and spruce, with some larch. Usually, the types of tree that grow at any level are determined by the soils. Oaks, limes, hornbeams, and spruce like the deeper, richer soils. Beeches, chestnuts, and pine like the thinner, more acid soils, and the ash likes soil with lots of lime in it.

Coniferous trees are more suited to the conditions of the higher levels of the mountains because they are adapted to survive with a shorter growing season. For instance, it can take two seasons for the cones to develop and release their seeds. This suits the shorter growing season found at the colder heights. The soft wood of pine and spruce stores more water than that of hardwood trees, so drying winds in summer can be resisted. The stored water is also useful in spring: It allows growth to begin quickly when water in the ground is still too cold to move easily into the roots.

Since the wood of conifers is not as dense as that of deciduous trees, conifers do not have to make as much food in order to grow fast. So they are again well adapted to the shorter growing season. A coniferous tree usually takes only 90 to 100 years to reach maturity. A broadleaved tree may take 150 years.

At the highest levels in the mountains, above the tree line, are the alpine pastures. The plants that thrive here are not only well adapted to the high radiation (see page 8), but they also have other features that help them to survive at high altitudes. Bulb plants such as the crocus can grow beneath the snow, when there is sufficient warmth trapped in the soil by the snow blanket.

▽ Scrubby maquis vegetation on a Greek island.

They can bloom within a day or so of the snow's melting, using food stored from the previous season.

Other plants, such as the gentian, have deep green leaves to absorb heat, and some, like the edelweiss, have hairy leaves to cut down water loss in the dry summer winds. Some plants may wait for a number of years before they have sufficient stored food to produce flowers and seeds. Nearly all of them have extensive root systems to soak up water.

In early summer in the Alps and Jura mountains, many of the meadows above the tree line are a blaze of color. The varied foliage from all the flowers adds to the high quality of the summer grazing and the hay crop. Many of the leaves are rich in vitamins and minerals, and are very nutritious for the animals. The

Gentians (top), edelweiss (center) and orchids (below) flourish in the high-level meadows in the Alps.

milk from the cows grazed on these pastures is famous for its flavor and richness.

Sad to say, a number of the alpine species of the Alps, Jura, and other central European mountains are becoming rare. The mountain environment is quite delicate, and as the pastures are opened up by roads and ski lifts for summer and winter visitors, the soil and its plants are being damaged. The countries of the Alps—Switzerland, France, Italy, and Austria—are very eager to preserve the pastures. There are strong penalties for those who remove or damage the plants. However, much of the damage is caused by ski developments, which the local people want. (See page 39.)

The tree line

From north to south in Europe the level reached by trees in the mountains gets higher. In Scotland trees grow only up to about 1,600 feet. In Norway in the deep, sheltered, inner sea inlets (fiords) the tree line may rise to 2,000 feet. In middle Europe in Germany, France, southern Poland, Czechoslovakia, and Romania, the tree line reaches over 6,500 feet. In the central Alps it climbs to over 8,000 feet. However, in the Alps, the tree line shows an effect that can be seen in most of the world's largest mountain ranges. The tree line in the central highest parts is higher than the tree line in surrounding mountains. This is not easy to understand, as it is colder in winter in the higher mountains. However, in summer, it is actually warmer in the highest land, because then the central parts of the mountains have fewer clouds than the lower mountains around them. The rain-bearing winds frequently dry out by the time they reach these higher parts, so there are fewer clouds. The climate is sunnier and warmer, so trees are able to grow.

At the highest level of the trees—at the tree line itself—the trees may show the drastic effects of wind. They may have branches only on one side, as the buds on the side facing the strongest wind may be killed off. Sometimes the trees may never grow above a ground-hugging, gnarled, bushlike form.

Upland birds

In the mountains of northwestern Europe, many of the mammals are similar to animals of the surrounding lowland. This is not surprising since only 10,000 years ago the mountains and much of the lowland in Britain and Scandinavia were buried under a great ice sheet. Since that time there has not been a long period for a new population of animals adapted to mountain conditions to evolve. In Britain only one animal, the red grouse, has evolved to live only in the mountains. In fact, in Europe the most distinctive animals to depend on mountain ecology are the birds.

Like most animals, birds have evolved over time to be adapted to a particular sort of ecology. Their size and plumage, their feeding, nesting, flight patterns, and so on all reflect the kinds of environment to which they have become adapted. However, birds often use a number of different types of environment; many migrant birds, for example, live in different parts of the world from season to season. Birds also learn to live quite successfully in new environments created by people. The herring gull, for example, regularly moves inland from the coast to feed on garbage dumps.

The main division among the European mountain birds is between those that like closed forests and those that use the open moorlands and alpine pastures. In the first group there are some species that prefer deciduous woods and some that prefer coniferous woods, but most show no preference for one or the other. Not many of this group are confined to the mountains only; most are found in lowlands as well as mountain forests. The common treecreeper, for example, lives at any level in lowlands or mountain forests and shows no preference for deciduous or coniferous woods.

The short-toed treecreeper, on the other hand, has a definite preference for deciduous woods. Where its range overlaps that of the common treecreeper, the two species tend to feed in different areas. The common species usually lives at a higher level than the short-toed species. So in southern Europe it is more a mountain bird

△ A male red grouse

A young treecreeper after its first flight

△ Golden eagle chicks

than a lowland bird. By having different feeding grounds the two species do not come into conflict for the same food sources.

Many of the birds migrate between higher and lower levels from summer to winter. The citril finch of the Alps, Pyrenees, and northwest Spain prefers mountain coniferous woods in summer but migrates to the lower deciduous wood in winter. The snow finch, widespread in the highest mountains in Europe, lives above the tree line in summer but migrates to the valley forests in winter. Some of these birds, such as the alpine chough (a type of crow), have been attracted by the warmth, shelter, and extra food provided by the villages and farms.

Some birds have become upland species because their original range has been reduced. This is true of some of the birds of prey. In Britain, the red kite is now found only in a small area in the mountains of central Wales. Once its range stretched across Britain (it is still common in southern and central Europe), but large numbers were shot because they took farmers' chicks. At the turn of the century there were only a few individuals left. These were given protection and its numbers are slowly recovering—there are now 50 to 60 pairs—but it is likely that in Britain the red kite will now always be a mountain bird.

The same may have happened to the golden eagle a long time ago. In Britain

In Britain, the red kite is a very rare bird of prey found only in some Welsh mountains. In southern Europe, it is much more common and may be seen circling over roads, waiting to feed on animals killed by traffic. In flight, the red kite is easily identified by its deeply forked tail.

△ The chamois lives above the tree line in high mountains in Europe.

and southern Europe it is completely a mountain bird. But in the eastern part of its range, in the USSR, it is a bird of the coniferous forests. In winter it hunts over open fields. One suggestion is that it has been hunted out of lowland areas in western Europe, but has survived in the Russian lowlands because they are only thinly populated.

The birds of Europe that are clearly mountain species are those that live above the tree line in the alpine zone. Some of these birds are also found at sea level in the Arctic, where the tundra lands provide similar country in which to nest and feed. These birds include the Lapland bunting, the snow bunting, and snowy owl. They are also found on upland moors and fells in Norway, Sweden, and Britain. The ptarmigan lives in the Arctic tundra and in the alpine pastures of southern Europe. It changes its plumage to white in winter, a useful camouflage for a bird that tends to stay close to the ground. It makes burrows in the snow to survive the intense winter cold of these two environments.

Mountain mammals

Hunted down

The larger mammals of the European mountains are the remnants of once widespread populations of animals which colonized Europe after the Ice Age. Over the thousands of years since then, these animals have been hunted to a few remaining pockets of the most difficult mountain country. In Italy there is a small population of gray wolves left in the central Apennine mountains. Once, this animal was the most widespread mammal in Europe, Asia, and North America—apart from human beings. Other examples include the bear, wild boar, lynx, and wild cat.

All of the larger mammals are now protected by the countries where they are found. But although conservation may be successful in Europe, the animals will never again have the range they once had.

Real mountaineers

The truest mountain dwellers among the European mammals are the chamois and the

◁ The alpine ibex is very sure-footed. It can move along on tiny footholds. If it finds its way blocked, it can rear up on its hind legs and turn around. These two young males are play-fighting on a narrow ledge.

▷ An adult male ibex

alpine ibex, sometimes called mountain goats. In fact, they belong to the goat-antelopes, an animal family that stands midway between the true goats and the antelopes. Both animals are at home in the highest rocky parts of the Alps, well beyond the tree line. This environment is one of the most difficult of all the world's landscapes in which to live. It is extremely cold in winter, has high winds, and often offers only tiny footholds among the rocks. What little food there is is tough and not very nourishing. Much of the food, in any case, is buried by snow for long periods during the winter. Like the true goats, the chamois and ibex have several stomachs, which help to extract more nourishment from the poor food.

The chamois is quite widespread in the high mountains of southern Europe. It migrates down to the forest edge in winter to seek food. The alpine ibex, on the other hand, is very shy and rarely leaves the highest land. Both animals have suffered from hunting, but of the two the alpine ibex has suffered the most. Its magnificent long horns—which can be over 30 inches long—were prized as decoration for hunting lodges. By 1870 only a few dozen animals were left in the Gran Paradiso mountains in the central Alps. Then King Victor Emanuel II of Italy passed a law to prevent further killing. He also created a natural reserve in the mountains. This forward-looking act of conservation worked. The population in the Alps today is around 10,000. Their future seems safe, although they are now disturbed by high-level skiers (see page 41).

deciduous—describes trees that lose all their leaves each year.
hardwood—the wood of trees such as oak and beech, as opposed to the softwood of coniferous trees.
coniferous—belonging to conifers, trees that bear cones and needlelike leaves.

The mountains of North America

△ Alpine flowers in the Gore Range, in Colorado.

▷ Redwoods, also called sequoias, in Sequoia National Park in California.

Unlike the mountains of Europe, the mountains of Canada, the United States, and Central America cover a much larger proportion of the continent. For instance, about half of the United States land area is high land. Some of it is flat plain but much of it is mountainous with steep slopes. The mountain ranges on the western side are the Western Cordilleras, which include the Rocky Mountains, the Cascades, the Sierras, and the Coast Ranges. The major mountain system on the eastern side is called the Appalachians. Over two thirds of Mexico is high land with mountain ranges (called sierras).

There are many different kinds of climate and ecology contained within these high lands. The northern and central ranges of the Western Cordilleras have abundant rainfall, but the southern ranges (including the mountains of northern Mexico) are dry, with deserts. The Appalachians, which are lower, have heavy rainfall. Here there are large areas of deciduous forests. The wet tropical mountains of southern Mexico to Panama are like the tropical mountain forest of the Andes in South America.

The Western Cordilleras

The Western Cordilleras are the domain of forest and alpine pastures in the north, with forest, grassland, and alpine pastures in the center and forest, scrub, and desert in the south. Compared to the forests of Europe, the forests of these western ranges in North America have a much greater variety of trees. This is mostly the result of the effects of the Ice Age in the two continents.

When the ice sheets grew, the trees began to live farther south, where the climate was still suitable for them. In America, there was plenty of land in the southern part of the continent for each type to colonize places where it could survive. In Europe, on the other hand, the Mediterranean Sea lay across the latitude where the different trees could have found suitable places to grow. Many of the trees, therefore, could not survive through the cold periods of the Ice Age.

Before the Ice Age, the forests of Europe, Asia, and North America were very similar. There were many shared tree (and animal) types common to the forests of the three areas. For instance, the redwoods of North America once grew widely in Europe and Asia. After the Ice Age they were confined to the mountains of the northwestern United States, where the climate is foggy and wet. Earlier in this century a small area of redwoods was discovered in the cloudy, wet mountains of western China. These American and Chinese redwoods are the last survivors of a population of trees that once ranged through the northern hemisphere.

Tiny traces

Ecologists learn about the history of redwood forests from the traces that the trees leave behind. The most important trace is the pollen they produce. Pollen is made of a very tough chemical that does not rot easily. Each plant type produces its own kind of pollen grain, which can be identified under a powerful microscope from patterns on its surface. When pollen falls into swamps, it is preserved among the rotted plant remains that form peat. By extracting the pollen and analyzing what kinds are present, a picture can be built up of the vegetation of the past. Plants are related to climate and by studying vegetation in this way we can build up a picture of how climate has changed. The same information will also help us to predict how vegetation may change in the future (see page 42).

Conifer kings

The forests of the Western Cordilleras, from New Mexico and California to Alaska, are some of the world's greatest reserves of coniferous trees. The climate is quite suitable in many regions for deciduous trees, yet there are few of these. Again, this is a result of the Ice Age. Because the land is so high, glaciers occurred widely in the Cordilleras. The tough conifers, like sitka spruce, Douglas fir, the western hemlock, the western cedar, and the redwoods, could survive quite close to the ice edge. As the glaciers melted, these trees quickly colonized the mountains. The deciduous trees could not colonize the land so easily because the icy climate had pushed them much farther away from the mountains.

Farther inland, and in the south, the climate of the United States is drier than on the coast. Here, pines such as the lodgepole

◁ Sitka spruce cones 60 days old

△ Sitka spruce cones 141 days old

△ Sitka spruce cones 293 days old

Most conifers complete their reproductive cycle in one season. Where the sitka spruce (left and above) is growing in mild conditions near sea level it produces seeds in a single season, but at higher, colder levels it may not release the seeds until the following year.

pine, ponderosa pine, and the larch are suited to the mountains. Many of the valley floors and much of the hilly country between the ranges are grassland prairies. In the southern end of the Cordilleras these grasslands give way to deserts, as happens in Nevada, Arizona and New Mexico.

The last of the high coniferous forests are seen in northern Mexico. This is one of the great frontiers in the world's ecology. Nothing like these forests is found anywhere to the south of this area. In the southern hemisphere other trees form dense forests—trees such as the snow gums of the Australian mountains and the southern beech of Chile, Tasmania, and New Zealand.

▽ The snow gum, a kind of eucalyptus, is an evergreen tree that forms forests in the mountains of Australia.

Forest giants

The conifers of these lands are some of the world's largest living things. The mature western cedar and western hemlock reach 300 feet, and many coastal redwoods have reached higher. The record height for this tree is 394 feet. Only some Australian eucalyptus trees have come close to matching these heights.

Walking in these forests is like walking in a great cathedral. The straight trunks carry their branches high, and shafts of sunlight filter through them to the damp, shady forest floor. The ground is soft with a thick peaty carpet of needles, which supports many kinds of fungi. If you turn the needles over there are usually many large wood ants, but the soil below is not especially rich and sometimes looks like gray ash. Scientists use the Russian name for this kind of soil, podzol. It is not very rich in plant food, which is another reason why deciduous trees are not found in large numbers—they do not flourish on this soil but conifers do.

Where sufficient light reaches the forest floor there is a rich growth of ferns, mosses, and a few flowering plants, including willow, alder, birch, aspen, and berry bushes such as juniper. Occasionally a deer can be seen scampering away as you approach. Not many birds can be seen but they can be heard high in the branches.

Following the trails in the forests may bring you suddenly to an area of cleared land. The scene is often one of devastation. Great trunks lie around where the foresters have left them. Old dirt roads cover the mountainsides. The soil is often partially washed away and deep gulleys may have formed. This scene is repeated all over the western United States. It is common in areas where, in the past, the forestry was not properly controlled.

Science to the rescue

The Forestry Service is now working toward much more scientific forestry methods in forests in the United States. Research has been carried out for many years to find the best ways of cutting the timber without damaging the ecology in a permanent way.

◁ Clearcutting (clearing all trees from an area) led to soil erosion, and has now been abandoned.

Before 1960, selective cutting was widespread; only the best trees were taken in patches. The remaining trees were often damaged by heavy vehicles or removed to make roads through to the selected trees. Between 1960 and 1980 clearcutting was practiced, by which all the trees were removed. This method led to soil erosion and poor scrubby regrowth.

Now the Forestry Service has shown that the best method seems to be to remove about 85 percent of all the trees, then leave the area alone for decades. The remaining trees provide new seedlings and the natural ecosystem re-establishes itself in time. The pest populations of insects and fungi never have the right conditions to become established, or they eliminate only the weakest seedlings. Forestry like this also helps to conserve the animal populations. By 1992 the Forestry Service will have a full research program to test these methods into the next century.

In other parts of the world, foresters have found the huge conifers of North America to be more productive than the native trees in their own lands.

Keeping time

The western mountain forests contain not only the world's largest trees, but also some of the oldest. Some of the redwood giants are over 2,000 years old. However, the oldest tree of all is the bristlecone pine of the southern Californian mountains. These pines grow at high levels in nearly rainless conditions. They are very slow growing. Some bristlecones are over 4,000 years old. Scientists have been able to use their tree rings to find out a great deal about the past.

The tree rings contain a radioactive chemical called carbon 14, which trees take from the air as part of the carbon dioxide they need to make their food. The amount of carbon 14 in the air varies from time to time. So each tree ring preserves the record of how much carbon 14 was in the atmosphere in the year when the ring formed. The rings in turn can be used to date old bone, cloth, and other materials, which also contain carbon 14. Tree rings with the same amount give us their approximate date.

This "tree-ring calendar" was discovered in the 1970s. It has shown that many theories about the past were wrong. For instance, it used to be thought that the great pyramids of Ancient Egypt were built before the stone monuments such as Stonehenge in Britain and Carnac in France. Nobody imagined that the prehistoric people of northern Europe were more advanced than the Ancient Egyptians. The new carbon dating shows these stone structures are, in fact, older than the pyramids!

▷ Bristlecone pines may live to be thousands of years old.

▽ The thick, fleshy needles of the bristlecone pine are ideal for low temperatures.

▽ Lodgepole pines ablaze in the 1988 fire in Yellowstone National Park in Wyoming.

Looking ahead

In 1988 the famous Yellowstone National Park in Wyoming suffered a huge fire. Tens of thousands of acres of forest were destroyed. Yet the fire services did little to save any of it. This was deliberate policy. Many of the people who visit this magnificent forest country found it difficult to understand why little was done to save one of the nation's great forest attractions. However, by doing little the National Park Service was recognizing that fire is a normal part of the ecology of the mountain forests. In fact, some species of trees, such as the black spruce, actually need fire. Their cones will not open unless they are scorched. In allowing the forest to burn naturally the authorities were following nature's pattern. In 100 years, the magnificent forests will be flourishing again.

Ecologists in the United States have set up research projects to monitor what is happening to a range of environments, on the scale of 100 years. The full results will not be seen until 2090. In future, it will be necessary to manage many other sites in the world on a similar, long-term time scale.

Animals of the mountains

As in Europe, many of the animals in the American mountains can also be found in the lowland areas. The deer, bear, and moose like forest cover. The pronghorn, marmot, and buffalo like open grassland. Animals that like the higher alpine areas include the elk, the bighorn sheep, and the Rocky Mountain goat (a type of chamois). These three animals are widespread in the Cordilleras. They migrate from the alpine lands in the winter, to the shelter of the forests. The bighorn sheep and the Rocky Mountain goat are very sure-footed, and can travel fast over rocky ground and almost sheer rock slopes with complete confidence. Their hooves are flexible and the toes act like pincers. The bottom of the hoof is arched and fairly soft.

The **cougar**, one of the main mountain predators, preys on deer, moose, bighorn sheep, and chamois, although the sheep and chamois are rarely caught in their steep, rocky summer territories. The cougar was once common throughout the continent. It is another example of an animal that has become a mountain creature because it can find refuge there.

The other main predator in the mountains is the gray wolf, whose numbers have been much reduced by hunting, except in Canada and Alaska. Once the wolf preyed on the forest herbivores, the deer and moose, and on the buffalo in the grassy lands. Some buffalo herds are protected, and their numbers controlled by shooting. There have been suggestions that the wolf should be re-introduced to areas such as Colorado so that it can control numbers of buffalo in protected herds. But wolves also like to eat cattle, so the local people did not think this was a good idea!

▽ Cougar

▽ Bighorn sheep

In many of the central and southern grassland and scrub mountain areas, the wild horse has become common. Because they compete with cattle for grazing, they are often rounded up and sold for pet meat.

The role of the rodent

In addition to the larger creatures, there are large populations of **rodents**, many varieties of which are not found in Europe. Some rodents are very important to the growth of plants. For instance, in the high alpine pastures there are no earthworms to turn over the soil. Yet the soil does not become waterlogged. This is because of animals such as the meadow vole, which remains active through the winter. Its burrowing turns over the soil quite quickly. Its tunnels form excellent drains and its droppings provide manure. The vole stays under the snow—in winter if it comes above the snow it can freeze to death in minutes.

One animal actually cuts and stores the pasture vegetation. This is the pika, which hides vegetation in the rock crevices in the summer to get ready for its winter feed. The remains of these stores are often colonized by plants and they eventually extend the soil-covered areas.

One important animal, the marmot (or prairie dog), is widespread on grasslands in protected prairies, and in rocky parts of mountains. It lives in large colonies, sometimes with hundreds of individuals, and can consume enormous amounts of grassy food. In many areas it is responsible for the failure of trees to establish themselves, since it eats off the roots of seedlings. In some areas farmers have poisoned large numbers of marmots to keep them from eating grass grown for cattle.

The future of the Cordilleras

The Cordilleras lie in developed countries with plenty of resources, so much is done to preserve and protect the wildlife of the area. The United States, Mexico, and Canada all have extensive areas of wild land protected by strong regulations. There are also many official parks, where access by people is limited. The needs of the wildlife in these magnificent mountains are kept in harmony with the needs of visitors.

cougar—also called puma, panther, and mountain lion.
rodents—small mammals with teeth for gnawing.

▽ Marmots in Alaska

Low latitude mountains

The Andes

When the Spaniards first explored the Andes and central America they very quickly saw that the climate and the vegetation of the mountains in the northern tropical part reflected the fall in temperature as height increased. The contrasts produced by this fall were so strong the Spaniards gave each zone a special name. These are: ***tierra caliente*** the hot, wet valley floors; ***tierra templada*** the warm, middle slopes of the valleys; ***tierra fria*** the cool, upper parts of the valleys.

Each of these zones has its own distinct range of plants and animals, which thrive in the particular climate of the region. The tierra caliente, below about 3,250 feet, is tropical rain forest. This is the haunt of the jaguar, cougar, tapir, tree sloth, and many kinds of monkeys, bats, and birds, reptiles, and amphibians. It is the zone for growing sugar cane, bananas, cacao and—on the valley sides—coffee, corn, tobacco, and coca.

The tierra templada, between 3,250 feet and 6,500 feet, has much less rainfall and is often covered by thin forest or scrub. It has some farming and grazing but not as much as the zones above or below it.

The tierra fria, above 8,200 feet, has good forest cover and is the zone for wheat, potatoes, and other vegetable and salad crops. Most of the people in the Andean region live in this zone. In Bolivia it is a high dry plain called Altiplano. Beyond this zone, there are extensive alpine grasslands with no trees, called the puna. This is the home of the mountain grazing animals: the vicuna, llama, alpaca, guanaco, and their main enemy, the cougar. Beyond this again is a bleak land of rocks with lichens and mosses, called the paramos, which extends to the snow line. The steep cliffs of this zone are the nesting sites of the huge Andean condor.

△ Andean condor

▷ Quechua Indian mule herders prepare potatoes (inset). They live in the high puna landscapes of the Andes (main picture).

High living

In some regions of the Andes in Peru and Bolivia, the climate of tierra fria is described as "perpetual springtime." Fruit trees can be in blossom at any time of the year. Although the nights are chilly, the daytime temperatures are warm and there is lots of sunshine. The fields can produce a succession of crops every year. The land is also above the zones where tropical diseases flourish. It was in these pleasant conditions about 4,000 years ago that villages and cities began to develop. The greatest of these early civilizations was that of the Incas. When the Spanish came they destroyed the Inca empire but the people remained. The crops and animals they farmed, and their languages such as Quechua are still the basis of the life in these regions.

Although the climate can be pleasant at these high levels, the thin air is a

Llamas (left) are thriving in the high pastures of the Andes, but the vicuna (above) is an endangered species.

disadvantage. Not only is there less oxygen to breathe but there is less air pressure to force the oxygen into the blood. Over the centuries the Andean people have become adapted to this. A person living at 13,000 feet on the Bolivian puna has around 20 percent more blood than a person living near sea level. The blood also has over half its volume made up of red blood cells (which carry oxygen). People from lowland areas have fewer red cells, and often get **mountain sickness** at these heights.

Endangered species

The alpaca, llama, guanaco, and vicuna, all grazing animals of the extensive puna pasture lands, are relatives of the camel, known as cameloids. Three of these species are widespread, but the fourth, the vicuna, is now one of the world's endangered species.

This small cameloid lives at the higher edges of the puna, and even into the paramos, driven there by farmers hunting it out of cattle grazing lands. Once the vicuna was widespread through the Andes, where it was protected by the Incas. Only the Inca ruler was allowed to wear its wool. Since the Spanish came it has been hunted for its fleece, which produces the finest and most expensive wool in the world. However, the animal is so small and the wool so fine it has never been considered worth domesticating.

For over 100 years now, attempts have been made to conserve the vicuna. A large reserve has been created for the animals in the Andes of southern Peru. This was largely through the efforts of one man, Felipe Benavides. He also showed it was possible to domesticate the vicuna for its wool. If it could be domesticated by the local people it would produce a new source of income for the poor farmers of the region. At present many of them grow coca for the illegal cocaine trade.

The Himalayan lands

Many of the problems of life in the Andes for people and animals are repeated in the other main low latitude mountain lands, in Africa and the Himalayas. The Himalayan lands form the world's largest and highest mountain mass. They include the Himalayas, the Pamirs, and Hindu Kush, which make up the great mountain wall of the Indian subcontinent. To the north and east of these are the ranges of the Karakoram, Kunlun Shan, Tien Shan, Altai, and those of western China.

In the southern ranges the ecology is dominated by the distribution of rain from the monsoon winds. These winds bring the rain from the Indian Ocean in summer. The eastern Himalayas have some of the highest rainfalls in the world. The western ranges, however, are much drier.

The zones of vegetation are best seen in the eastern Himalayas. In the lower parts of the mountain fringe are thick forests of mixed deciduous and evergreen trees such as sal and teak. Above these trees are more dense forests of mainly deciduous trees. At about 5,000 feet, forests consist of trees that like a cool, wet climate, including chestnut, maple, alder, birch, magnolia, and laurel. These give way to conifer and rhododendron forests from around 10,000–13,000 feet, where the alpine pastures take over. Finally the snow line is reached at around 16,000 feet. Throughout most of these vegetation zones (apart from the alpine zone) there are frequent bamboo thickets.

▽ These Himalayan peaks in Nepal are more than 21,000 feet high.

Animal geography

About 100 years ago, the German scientist P. L. Sclater suggested that the land animals of the world existed in definite groups. Each group more or less matched a continent—North America, South America, Africa, Australia, and Eurasia. This theory is still used as the basis of animal geography. Sclater divided the last group, Eurasia, into two parts: a northern region named the Palaearctic and a Southeast Asian region, the Oriental. The junction of these two regions is the Himalayas. Therefore, we find in the mountains a mixture of animals. Some, like the pandas and monkeys, are also found elsewhere in Southeast Asia. Some, like the ibex, are also found in Eurasia. Some animals, like the tiger, belong in both regions. Because of this junction of geographical zones, and the size and remoteness of the region, the Himalayan mountain fauna is the richest of all the world's mountain areas.

The animals from the northern cooler region, the Palaearctic, tend to live in the cooler, higher zones. The more Oriental animals tend to like the warmer, lower zones. Thus, in the alpine grassland we find the brown bear, the marmot, the pika, and four species of wild goat, including a true mountain goat, the markhor. There are also goat-antelopes like the ibex, the goral, the serow, and a number of varieties of wild sheep. One, the blue sheep, has features of both goats and sheep. There are also antelopes such as the chiru and goa, deer such as the muskdeer, and the cowlike yak.

◁ A yak driver and his yak, in Nepal

▽ Snow leopard

The yak is very important to the people of the Himalayas. It is a tough and valuable beast of burden. Its strength and sure-footedness, its ability to withstand cold and to go for long periods on little water or food and even to provide milk are the main reasons why people have been able to live in many of the high valleys. Above 5,600 feet most households of the villages in Nepal, for example, move from the valley floors to the alpine pastures and back again as the seasons change. This practice is called transhumance. At the lower levels families produce their crops and graze their yaks, goats, and sheep. In the summer, the families and their animals move to grazing lands on the high ridges. Many of the world's mountain people practice transhumance.

Below 6,500 feet, in the warm teak and sal forests, the animals are much more Oriental in type. They include the elephant, the rare Indian rhinoceros, and many deer varieties such as the muntjac and swamp deer. There are also a number of antelope and gazelle species, buffalo, monkeys, and wild pig.

The predators in the lower mountain forests include the leopard, tiger, hyena, and wild dog. Preying on the larger animals in the alpine zone are the magnificent snow leopard, the black bear, and the eastern Himalayan blue bear. The smaller animals are eaten by various cat species and the Tibetan weasel, and by birds of prey, especially the golden eagle. Carrion-eaters like the choughs, griffon vultures, ravens, and lammergeiers are widespread summer visitors through the alpine zones. These smaller birds often follow people and their animals. The alpine chough has been known to visit mountaineers' camps at over 26,000 feet. Birds that remain at high levels through the year include the Tibetan and Himalayan snow cocks and the snow partridge, which nest and breed above 16,000 feet.

Panda territory

Much of the mountain fauna is threatened by change in the environment. This is very obvious where the human population is growing quickly. In the bamboo forests of China, more and more people are moving in to clear the bamboo for farming. The bamboo thickets are reduced to islands of plants on the valley sides. But the giant panda needs large areas in which to roam. It does not breed easily without a large territory. This may be one of the reasons there have been few successes in zoos, where it is confined in a small space. As the bamboo islands get smaller, the chances of pandas meeting to breed lessen. Its numbers fall as a result. Because Asia has some of the fastest population growth in the world, careful planning will be needed to keep the wonderful animal variety of the mountains for future generations.

The Mountains of the Moon

In tropical Africa, volcanic mountains such as Mount Kenya, Kilimanjaro, and Ruwenzori stand high above the great plains, which cover much of the interior of the continent. These mountains, although not extensive, have one of the most distinctive ecologies of all the world's mountains. The Ruwenzori mountain mass, for instance, known as the "Mountains of the Moon," is only around 100 miles long yet contains five distinct life zones. It also contains some of the most difficult country to move in of any in the world.

The grasslands with trees (called savanna), which cover the plains surrounding Ruwenzori, extend almost up to 6,500 feet. They end abruptly in a dense rain forest of tall trees with huge foliage and tree ferns. This is the home of the forest buffalo, and there are some small villages. The rain forest vegetation continues for the next 3,000 feet, and then is replaced by dense bamboo forest. Bamboo can grow very fast and is extremely difficult to clear in this wet land. One study measured its growth at three feet a day and it can reach 100 feet in about eight weeks. There are no permanent villages in this zone, and few animals.

Beyond the bamboo, at around 10,000 feet, is the heath forest. This kind of vegetation is unique to the African mountains. The dominant plant is a tree heather related to the heather of the Scottish moors. But in this case the plant can be 30 feet or more high and may be festooned with lichens. The ground cover is a thick, soggy peaty bog, luxuriantly covered with mosses and liverwort. It is extremely difficult to walk on this surface. There are few large animals in this zone and the local people rarely go into it.

Finally, the alpine zone is reached at around 13,000 feet. It is nothing like the grassy alpine zones we have seen in other mountain ranges. Tall tree groundsel, heathers, lobelias, senecios, and large grass tussocks cover the surface. The land is almost permanently misty. Only toward the summits themselves does the land begin to resemble the familiar alpine zone. Here the rocks are covered with moss and lichens and there are isolated flowering plants. The plants grow up to the edge of the permanent snow at around 16,000 feet.

Vegetation similar to the Ruwenzori type is found on many of the other African volcanic peaks, such as Mount Kenya and Mount Kilimanjaro.

Many animals around the world find refuge in mountains when their living space is threatened, especially by people. In central Africa, the best example is the mountain gorilla. One colony lives over the border from Ruwenzori in the mountains of Rwanda. Here it lives in a protected national park with abundant natural forest and bamboo for feeding. The gorilla population in this colony had increased to over 300 by 1990. Unfortunately, in that year fighting broke out between revolutionaries and government troops. The noise terrified the animals and they began to flee wildly. Ecologists are very worried that the future of this harmless creature is no longer secure.

mountain sickness—nausea, headache, and shortness of breath caused by climbing to high altitudes.

△ Tree groundsel in the alpine zone of the Ruwenzori mountains
◁ A female mountain gorilla with its infant
▽ Mosses and lichens in a heath forest of the Ruwenzori mountains

The threats to the heights

In the less populated parts of the world the mountain ecology is not yet under great pressure. But where populations are large, or growing, there are conflicts between the ecology of plants and animals and the needs of people. The main threats are:

Deforestation for farmland. This happens widely in the Andes, Central America, Africa, India, Nepal, Southeast Asia, and China. Deforestation can also occur as people use wood for fuel or for building.

Deforestation for lumber production. This is widespread in Southeast Asia (Indonesia, Burma, the island of Borneo, Thailand, and southwest China) for tropical hardwood trees like teak and sal.

Development of dams for power production and water storage. Wildlife is affected by dam-building in rich countries, such as France, Tasmania, the United States, and Canada, as well as in poor countries such as India and China.

Mining and quarrying developments. Mountain rocks are often laced with mineral veins. Mountain mining can present a threat to wildlife in almost every country.

Tourism, including skiing and climbing. The threat here is mainly in the rich countries, but areas such as the Himalayas and the Andes are also under pressure to open new roads for tourist access.

Possible changes in the atmosphere and the world's climate.

About 30 years ago almost no attention was paid to any of these threats to the mountain ecology. Power projects and dam building, mining and quarrying, road building, skiing and hotel building were looked on as bringing benefits to remote and poor areas. Damage to wildlife from these activities often did not show up immediately. When it did, people in many parts of the

△ A copper mine in Chile

▽ A mountain road in Norway

world began to unite to control these developments or even to stop them altogether. In France, a plan to dam the upper part of the Loire River in the Auvergne mountains for power production was defeated in 1991 after several years of protests on the site and in courts of law. In Tasmania, the most untouched area of the island in the southwest mountains was saved from flooding caused by dam building. In India, the beautiful forested Doon valley, in the foothills of the Himalayas, was saved in 1988 by a decision of the Supreme Court to stop all quarrying. The ravaged forests are being replanted.

Victories like these are good news for the world's mountain ecology, but there are many areas where the damage is very serious. Correcting this damage will take many years. This is particularly true of places such as the Alps, where enormous sums of money have been invested in tourism, especially skiing.

The downhill side of skiing

In the Alps as a whole there are now over 40,000 ski runs, and more are being built. The runs are served by complexes of lifts, tows, new roads, and hotels. These ski stations attract millions of visitors each winter. (About 120 million tourists visit the Alps annually.) Building work on this scale means that the soil is damaged. In the summer, after the snow, these places are a dismal scene with large areas of bare soil and ruined plant cover.

Many of the skiers come by car, so new roads have to be built in the forests to reach the higher hotels and ski runs. In the Ziller valley, near Innsbruck in Austria, at the peak season over 20,000 cars use the narrow road up the valley. At present conservationists are trying to prevent the building of a new highway through the forest. Another Austrian valley, the Öztal, has about 7,000 permanent villagers but over the season two and a half million visitors arrive, mostly by car.

◁ A ski lift being built in Switzerland

With the car comes air pollution. The narrow, high sides of the valleys trap the exhaust fumes from the cars. The thin air allows the bright sunshine to produce chemical reactions between the car exhaust gases. The new chemicals then poison the trees. More than half the trees on many of the northern alpine slopes are dying. These sick trees can no longer give protection against avalanches, which are therefore becoming more frequent. As tree growth is slow at these heights, it takes a long time for new trees to replace the old. Great stripes of bare rock where avalanches have stripped away the trees and soil are now a familiar scene in most Alpine valleys. Occasionally, whole villages that have been protected by trees for centuries have been destroyed.

Even the areas away from the ski runs are not safe. Helicopters can reach the highest levels, where skiers disturb the rare birds and animals. In the valleys below, the alpine rivers and lakes have become increasingly polluted by sewage waste.

Most authorities are agreed that the ecology of the Alps has almost reached breaking point. At present, very little of the money made by the tourist industry is put back to conserve the very landscape the visitors come to enjoy. Up until now there has been no real planning for conservation, but a longer-term view is needed if Alpine wildlife is to survive.

Helping the Himalayas

Felling and flooding

So far, the Himalayan mountains in Nepal have not suffered as the Alps have. However, the growth of population as disease is brought under control is affecting the forest resources. The villagers of Nepal use the trees for constructing their homes, and they cut a great deal of wood, especially oak, for roof tiles (which are replaced every five or six years) and bamboo for storage baskets. As most Nepalese have a winter and a summer home, they need about 18 cubic feet of wood per person per year for these uses. A household also needs about 1,650 lb of wood fuel at the lower levels of the valleys and 4,400 lb at the cold high levels, for cooking and heating. As population grows, the numbers of trees that are cut down grows as well. Most of the wood that is needed is hardwood from trees such as oak, which are slow-growing.

The loss of trees opens up the soils of the mountainsides to the heavy monsoon rains and the soil begins to wash away. The diagram on this page shows how the amount of sand, pebbles, and other sediments carried by the rivers has increased. The rivers are also swollen by floodwater in the rainy season because the trees no longer control its flow. This can produce disastrous flooding in the plains of northern India, away from the mountains.

There are now many projects in Nepal and India to replace the lost wood by planting new forests and to preserve those that are left. In this, the local people are helped by the government and by foreign aid organizations. There are also plans to reduce the use of wood for cooking by employing solar energy. Eventually, the forests will provide opportunities for new local lumber and craft industries, which will be run on a small scale. Large-scale forest industries are not being encouraged to develop further.

Ecotourism

The conservation of mountain animals does not mean simply keeping them in zoos or small protected areas. Many mountain

A RIVER'S BURDEN

The bar graph shows the quantity of mud, sand, and pebbles that is carried by the main river in Nepal. Note the steady rise in the amount of sediment as mountain forests are removed and soil is washed into the river.

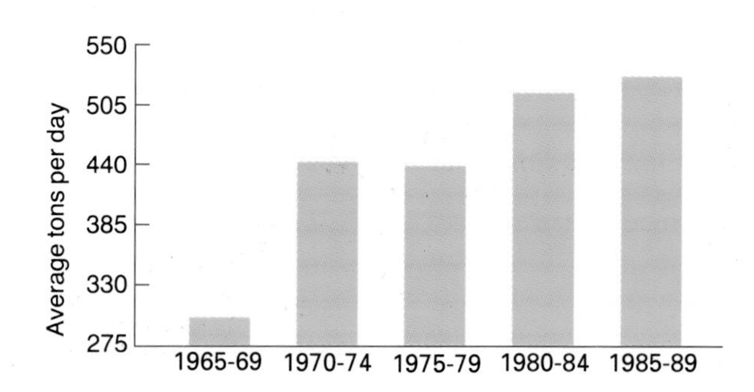

△ Trees have been cleared from the hillside behind this village in Nepal. The wood is used for building, cooking, and heating.

animals need large territories if they are to feed and breed regularly. In order to protect their territories governments may have to set aside large tracts of land and not allow development for mining, logging, farming, dam building, industry, or recreation. But these activities can be profitable, so a decision to limit them may not be very popular in poor countries.

One answer to the problem is to use the plants and animals as a source of income. The mountain gorillas of Rwanda, for example, have attracted many tourists. Many poor countries see ecotourism as something that will grow in the future. Ecotourism vacations combine adventure, learning, and conservation. They bring wealth to poor countries and provide much needed employment. Of course, if too many people come, then this form of tourism may destroy the very environment it aims to protect.

Keeping track

The starting point for protecting wild animals is to obtain the maximum amount of information on what the animals' needs are, (often called base-line information) and about what is happening to affect those needs. Only then can we take proper steps to help the animals. Around the world this information is being gathered by many universities, governments, and organizations such as the World Wide Fund for Nature (WWFN). The WWFN keeps a close watch on endangered species and has a "Red Data Book," which lists the species most in danger. Many of these are in the mountain areas. For instance, in the cool wet beech forests of the New Zealand mountains is one of the world's rarest animals. This is a ground-living parrot called the kakapo. Once it was widespread but now only a few remain. The bird has been given unlimited protection, which means that not even tourists are allowed in the area where it lives. The WWFN provides money to fund research into and management of its habitat.

Photography by space satellite is now being used to help in ecology management. The U.S. Landsat orbiter, for instance, can provide rapid, cheap photographs of vegetation. These pictures enable people to make accurate maps, which would be difficult as well as time-consuming to draw up from surveys on the ground. Any changes in the ecology of an area can be seen and steps taken to manage the change before damage is done. However, there are changes going on in the world's ecology that can be managed only on a scale of the whole earth itself.

Global changes

Inland from the port of Santa Marta in northern Colombia is a mountain mass which rises to about 20,000 feet. This is the Cristobal Colon. The people who live on its slopes have followed the same way of life for hundreds of years. They do not trade, and they have little contact with the world beyond their villages. In 1989 they invited visitors from the outside world to their homeland for the first time, to warn them that serious changes were taking place in their land.

The Cristobal Colon is high enough to have snow on its summits. The local people use this snow as a reliable source of fresh water for drinking, and as it melts it waters the vegetation, which would otherwise dry up. The villagers noticed that the snow patches were becoming smaller year by year. The vegetation of the mountain was becoming thin, and dying. The people of the mountain are convinced that these changes are caused by pollution from the world outside their mountain home.

Many scientists agree with the people of Cristobal Colon. The two main changes they see are that first, the atmosphere is becoming warmer, and second, a high layer within it, called the ozone layer, is getting thinner. Both these changes may be caused by gases we are producing in our modern way of life.

The warming of the air is called the greenhouse effect. Gases such as carbon dioxide, methane, and nitrogen and oxygen combinations absorb heat radiated from the land. They prevent it from escaping to space, so heat is kept in the air like heat in a greenhouse. These gases come from the burning of fossil fuels from chemical industries and so on. They may also come

△ The kakapo parrot from New Zealand (above) and the bald eagle of North America (right) are two threatened species of mountain birds.

▽ The langur monkey is one of very few monkeys that can live in the mountains of Asia.

from the use of chemical fertilizers in farming, from rotting vegetation, and even from the belching of cattle.

Not all scientists think that the greenhouse effect is real. However, there is no doubt that the earth's air is heating up. We do not yet know what all the effects will be, but the sea level may rise as ice melts in mountain glaciers. Plants and animals at high levels may actually benefit if the climate becomes less harsh. On the other hand, if the mountain climate is milder, then forest trees may invade the alpine pastures.

Ozone is a gas that absorbs ultraviolet rays from the sun. Ultraviolet (UV) light is very damaging to living things. The ozone layer, high in the atmosphere, is destroyed by a group of gases called chlorofluorocarbons (CFCs). CFCs have been widely used in domestic goods. As the ozone is reduced, the plants at high alpine levels and even the forest trees may be badly affected by the extra UV light that gets through the atmosphere.

The world may have realized the danger from CFCs in time. In 1987 the nations of the world signed an agreement called the Montreal Protocol to stop their use. Unfortunately, the chemicals are long-lived, so those already released will still be affecting the atmosphere for many years to come.

As we have seen, the ecology of the world's mountains is under threat in many ways. However, we now know more about what is happening, so we are in a better position to protect these environments. In the next decade there are likely to be many more agreements like the Montreal Protocol. Such agreements will help to protect many of the world's environments, including mountains, and allow them to endure into the future without too much further damage.

▽ Guanacos in Patagonia, Chile (below), and bald eagles in Alaska (see page 43) live at opposite ends of the western mountains of the Americas. But both are threatened by changes in their environment, caused by pollution, human activities, and possible future changes in world climate.

Index